MY HOME IN THE UNIVERSE

POEMS AND PLAYS FOR SPACE SCIENCE

To Tara Oaks Elementary

Brod Bagert 2019

Brod Bagert's HeART of Science
www.brodbagertsheartofscience.com

Introduction—Old Friends in the Night Sky

And so it's—Here-I-go-again!
I'll search and search until the end,
to understand things through and through,
'cause learning stuff is what I do.
from Hungry Brain

I decided to start with this quote from *Hungry Brain* so I could talk to you about learning. (This might turn out to be an important thing for you to hear, so please stay with me.) Learning is not just something you do to get good grades in school so you can graduate and get a job. Learning is something you do because knowing stuff makes life more interesting, and it's fun. You may think I'm crazy, but if I had a choice between being a learner and being a billionaire, I would definitely pick being a learner, and I'm going to try to explain why with a story.

When I was a little boy, my dad taught me how to find the constellation Orion. It was a chilly November night, and we were looking at the stars as we walked on the planks of a small boardwalk over Bay St. Louis in Mississippi. Halfway to the end he stopped, knelt next to me, and pointed his finger toward three bright stars in a perfect line. "That's the belt of the great hunter, Orion," he said. (My father knew a lot about the stars from being a navigator on a warship during World War II.) Next he pointed out the stars that represented Orion's club hanging from the belt, then his shoulders and feet. He told me the name of each of the stars, but the only two I remember are Betelgeuse (because it's red and has a funny name) and Rigel (because it's blue and very bright).

At the time I didn't think much about it, but for the rest of my life when the cool nights of autumn finally arrived, I always found myself turning my eyes toward the southern sky to find "the great hunter, Orion." And he never let me down. Like an old friend, he was always there. Eventually I learned to find a few more constellations and memorized the names of some of the brightest stars. Then I got curious about that big cloud of stars that someone told me was the Milky Way Galaxy, and it went on from there. The more I learned the more I wanted to learn.

I didn't become an astronomer and there's a ton of astronomy I don't know, but for me it's not about studying something and passing a test. It's about continuing to learn, little by little, for the rest of my life. Because now, when I look at the night sky, it's not just a bunch of sparkling lights against a black background. It's a place were I find old friends and cool stuff I can picture in my head—a giant nebula cloud of glowing gas zillions of miles from one end to the other, a big black hole sucking up everything that comes within its event horizon, an outcast star that isn't part of any galaxy and just soars though space all by itself. Becoming a learner makes life tons more interesting, and the best part is that you get to pick what you want to learn. Going to school is just a way of getting you started, but after that you get to keep learning on your own, and what you choose to learn is completely up to you.

So all of what I just said is why I wrote this book. For most of you it'll be a fun way to learn the basics of outer space. For some of you—future astronomers, rocket engineers, astronauts—it may be the beginning of a lifelong journey. So get ready because space is full of amazing stuff.

TABLE OF CONTENTS

Countdown Jitters........1
Principle of Rocket Engines, Newton's Third Law of Motion.
Sky Pictures........3
History of Constellations.
Little Dipper........5
North Star, Binary Stars, Axial Precession.
Around the World........6
Zero-G—Weightlessness in Outer Space.
Children of the Sun........7
Planets, Their Discovery, How We Lost Pluto.
Ride a Rocket........9
Planets, Asteroid Belt, Kuiper Belt, Oort Cloud.
Hungry Brain........11
Habitual Learning, Heliosphere, Termination Shock, Oort Cloud, Astronomical Units, Light Years.
Inner Heat and Outer Cold........13
Inner and Outer Planets.
My Lying Eyes and Sun Center........15
Geocentric and Heliocentric Models of the Solar System.
A Superhero Revolution........17
Aristotle, Ptolemy, Copernicus, Kepler, Galileo, Newton.
Make a Wish........19
Meteoroid, Meteor, Meteorite, Friction and Heat.
A Nice Place to Visit........21
Moon: Mass, Gravity, Weight; Atmospheric Pressure, Boiling Temperatures.
Peek-a-Boo Moon........24
Phases of the Moon, Waxing, Waning.
Turnabout is Fair Play........25
Lunar Eclipse, Solar Eclipse.
Ring Around the Sun........31
Venus both Morning Star and Evening Star, Transit of Venus.
Sunshine Science........33
Composition of the Sun, Hydrogen Fusion Process.
A Nerd Love Poem........35
Kinds of Stars, Black Holes, Age of Light from Stars.
Black Hole Brain........37
Life Cycle of Stars, Singularity, Black Holes, Neutron Stars, White Dwarfs.
Biggest?........39
Galaxies: Size, Contents.
The Wall........41
The Paradox of Endlessness, Hubble Deep Field Image, Distribution of Galaxies.
Alien Eyes?........42
Odds of Extraterrestrial Life.
My Home in the Universe........44
Post-Copernican Angst.
You Will Find Me........45
Unnamed Stars, Celestial Coordinates.
An Invitation to You........46

Countdown Jitters

Ten…Nine…Eight…
Wait!

Seven…Six…Five…Four…
You count real well but don't count more.

Three…Two…
What can I do?

One…Blast Off!
Aaaaaaaaaaaaaaaaaaaaaaaah!
Ooooooooooooooooooooooo!
Taking off in a rocket is scary to do.

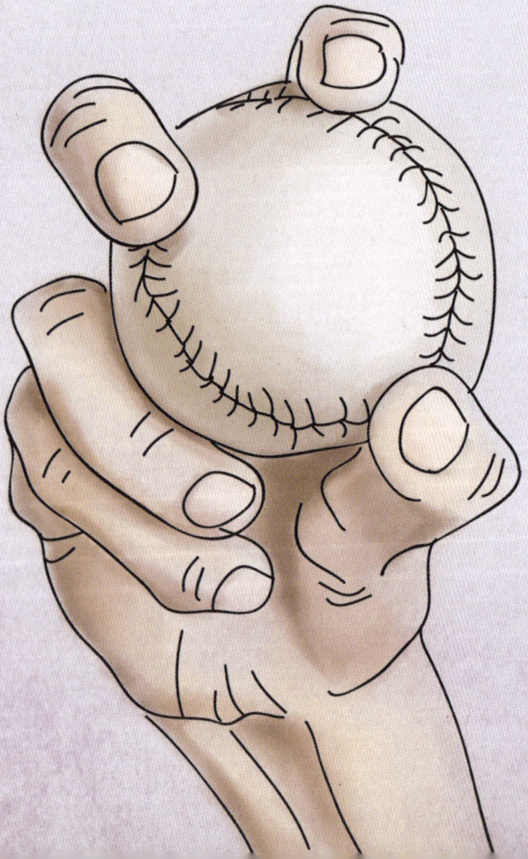

Everybody thinks rocket engines are really complicated, and in a way they are, but what makes them work is actually very simple. Inside a rocket, liquid fuel is converted into hot gases (it burns), and the hot gases shoot out of the rocket's rear end which pushes the rocket in the opposite direction.

(Please turn to the person next to you and say this next sentence as though you're a scientific genius, and don't worry if you don't understand it yet.) "Ah! Newton's third law of motion: *For every action there's an equal and opposite reaction.*" (Very cool.)

So here's my favorite way to explain how rockets work. Imagine you're sitting on the back porch of a house floating in outer space. Next to you is a bucket of baseballs. You pick up a baseball and throw it away from the house as hard as you can. What happens? You and the house move a teeny-weeny bit in the opposite direction of the baseball. Why? Because *for every action there's an equal and opposite reaction*. When you start throwing those baseballs you become a rocket engine, and the baseballs become rocket fuel.

Sky Pictures
or
My Place Among the Stars

These constellations in the sky
seem very strange to me
'cause all the stuff they say is there
I never really see.

Pegasus is a horse with wings,
but all I see is a square.
If someone didn't point it out
I'd never know it's there.

And then for Taurus the Bull there's just
a big-ole giant V.
For Leo the Lion a question mark
is all I ever see.

The Little Dipper's better 'cause
the dipper's really there,
but then somebody tells you it's
supposed to be a bear.

I guess those guys in ancient times
who named the constellations
were all a bunch of dreamers with
some wild imaginations.

And if they can do it, so can I—
Oh look! What's that I see?
A brand new constellation!
I think I'll name it ME!

Constellations are groups of stars that form imaginary patterns in the sky. As you probably already know, you have to use your imagination because constellations don't look a whole lot like what they're supposed to be. There are a total of 88 officially recognized constellations, 48 of which were known in ancient times.

How ancient are they? The ancient 48 come to us through the Greeks some 2,500 years ago, and a lot of what the Greeks knew probably came from the Babylonians who compiled the first star catalogue over 4,000 years ago, but knowledge of the stars and the practice of naming constellations goes back even further.

Little Dipper

Little Dipper in the sky,
Little Dipper, please don't cry.
Not as bright as all the rest?
Brightness doesn't make them best.

All the others move around,
always on the roam.
You stay right there in the North
and point the way back home.

Little Dipper in the sky,
Little Dipper, please don't cry.
Small you are yet still you stay,
and that is how I find my way.

Polaris, the last star in the handle of the Little Dipper, is a very famous star. It's what astronomers call a binary star, which means it is actually two stars that revolve around each other. Even so, Polaris is not big and bright like the other famous stars. Polaris is famous because it's always in the same place, due north, which is why we also call it the *North Star*.

Now this is really cool. Being the North Star is a temporary job. It's a cycle that takes about 25,000 years. So in its present position, little Polaris is the North Star, but 5,000 years ago Alpha Draconis was the North Star, and 13,000 years from now Vega will be the North Star.

This cycle is caused by a thing called *axial precession* or the *precession of the equinoxes*. (Note: the word is *precession* with an "e" not procession with an "o.") It's the third thing scientists discovered about Earth's motion. First we learned that Earth *rotates on its axis* causing night and day. Second we learned that it orbits the Sun, each *orbit* making a year. And third we learned about *axial precession*, which is a lot more complicated than the other two, so I'm going to give you an oversimplified explanation to introduce you to the idea.

Earth rotates on its axis like a spinning top. The gravity of other bodies in our solar system, mostly the Sun and the Moon, cause the Earth to do a very slow wobble that makes a full circle every 25,772 years, which is what we call *axial* precession and why being the North Star is a temporary job.

Around the World

Recite as a hand-clapping chant

If you go into space in a giant rocket
please pack a yo-yo in your spacesuit pocket.
    All us kids down here on Earth
    would really like to see,
    an astronaut do yo-yo tricks
    afloat in Zero-G.

Sooooo...
If you go into space in a giant rocket
please pack a yo-yo in your spacesuit pocket.

    Space, space where I want to be,
    floating around in Zero-G.

    Space, space where I want to be,
    floating around in Zero-G.

    Space, space where I want to be,
    floating around in...

STOP IT!
...
...
... *Zero-G*

A scientist read this poem and explained to me that "Zero-G" is short for "Zero G force," which means the feeling or sensation of weightlessness. He explained that there is at least a little gravity everywhere, so there really is no such thing as zero gravity. If you traveled half way to the moon, Earth's gravity would still be pulling on you, but it would be very weak, about a thousandth of the pull at Earth's surface, and yes you would float around in your spaceship.

But in low earth orbit, Earth's gravity is almost as strong as it is on the surface. So why do astronauts float when they're orbiting Earth on the space station? My friend explained that "people on the space station are falling 'around' the Earth at the same rate as the Earth's surface curves away, so the height above the surface does not change."

So either way, the most important thing to remember is that if you ever become an astronaut please remember to "pack a yo-yo in your spacesuit pocket!"

Children of the Sun

Mercury's small, almost nothing at all.
Venus is bright and near.
Earth is a planet with deep blue seas
and a sky that's blue and clear.

Mars is red and angry.
Jupiter has an eye.
Saturn has rings of ice and stone
that circle round its sky.

Uranus and *Neptune*
are far away and cold.
So there, I know my planets,
and I'm only eight years old.

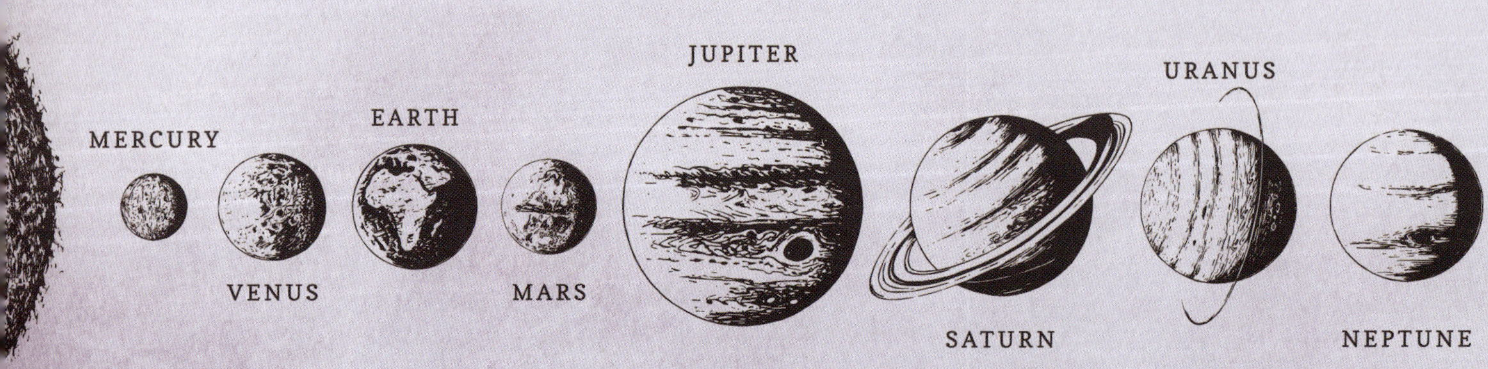

Mercury, Venus, Mars, Jupiter, and Saturn are the planets closest to Earth and can be seen with the naked eye (without a telescope), which is why astronomers have known about them for thousands of years. Uranus and Neptune, however, can only be observed through a telescope and had to be discovered.

Uranus was discovered in 1781 by Sir William Herschel. (Herschel's sister, Caroline Herschel, assisted him in his work and became a famous astronomer in her own right, which was extremely rare for a woman back in the 18th century. I often wonder if Caroline would have gotten credit as a co-discoverer of Uranus had she been Herschel's brother rather than his sister. You might want to do a little research and see what you think.)

Neptune was officially discovered in 1846, but, before it was actually observed, its existence and near-exact location were predicted independently by two astronomers—John Couch Adams and Urbain Le Verrier, both of whom share credit for the discovery.

But what about Pluto? I'm sure most of you have heard about poor little Pluto. It was discovered by astronomer Clyde Tombaugh and declared a planet in 1930. Then, in 2006, the International Astronomical Union downgraded it to something they called a dwarf planet. The problem was that in the Kuiper belt, which is where Pluto orbits, there are lots of large objects, some larger than Pluto. Eventually it was decided that for an object to be considered a planet it had to be large enough to claim its own orbit and kick out its competitors. And that is why Pluto is no longer a planet.

Ride a Rocket

Ride a rocket! Up we go!
Goodbye raindrops! Goodbye snow!
Earth has been a cozy place,
but we are bound for outer space.

Hear a roaring-rocket sound,
slowly lifting off the ground,
going faster, going high,
flying up above the sky.

Holy cow! How can it be?
Look I'm floating! Look at me!
Floating in a space cocoon
and now we're flying past the Moon.

From here we see a zillion stars,
but where-oh-where is planet Mars?
There it is, that big red ball.
Up close it doesn't look so small.

This dark and empty outer space
is such a cold and lonely place.
The coldest cold you've ever felt.
Watch out! Here comes the Asteroid Belt!

And up ahead, a glowing mass?
A giant orange ball of gas?
Dozens of moons! A humungous thing!
It's Jupiter, the planet king!

Saturn's rings! It might be fun
to make a little orbit run,
but we won't stop and make a fuss
'cause now it's on to Uranus.

First Uranus, then Neptune too,
both of them so big and blue.
Frozen balls of gas and stone,
so far away and so alone.

And next, the Kuiper belt I see,
this giant space in front of me,
this space where frozen objects twirl
here near the edge of solar world.

But that does not conclude our task.
"On no? What's left?" I hear you ask.
The Oort cloud? NO! Too far to roam.
Let's turn around and head for home.

I think the Kuiper belt is very exciting. Back when I was in elementary school, we didn't learn anything about it because no one knew it even existed. After the discovery of Pluto in 1930, astronomers began to speculate that there might be other objects orbiting beyond Uranus, but it wasn't until 1992 that the area officially became known as the Kuiper belt. Astronomers are still working to understand exactly how this area came to be in the first place and what's going-on out there in present time, which means that in the next few decades we're about to learn a bunch of NEW STUFF, which is why it's so exciting.

Now if that last sentence stirred your curiosity, I have a suggestion you may want to consider. Start learning about the Kuiper belt. Make it a little hobby. Learn the vocabulary. Read everything you can about it and keep an eye out for new discoveries. I think you're going to be surprised at how much fun this can be, and, who knows, you may someday work with other brilliant minds of your generation to increase our understanding. Yes, YOU have a brilliant mind. Normally functioning human brains are naturally brilliant. Then when you couple brilliance with the habit of curiosity, the sky's the limit. (The sky's the limit? Sky? A little pun? Okay, I promise not to do it any more.)

Hungry Brain

I'm a little embarrassed,
but I can't stop thinking about the Oort cloud.
Two weeks ago I didn't even know it existed,
and now I'm reading every Oort cloud thing I can find.
Last night it even talked to me in a dream.
"I am the Oort cloud," it growled in my head.
"Young earthling, have no fear.
I'm only a cloud of icy objects
beyond the heliosphere."
Please don't laugh,
I can't help it.

They think it's the source of all the long-period comets
that take hundreds of years
to come back through our solar system.
And they think it used to be closer to the sun
and got pushed away when the big planets formed,
and now it's so far away no one has ever seen it directly,
so it's all a big theory,
so I wonder—
Oort cloud, are you really there?
Some people say I shouldn't care.
"Five million-million miles away?
Too far to care," is what they say.

And yet, for me, that's just not right,
'cause learning is my heart's delight.
So when I think, "Dude, are you there?"
My hungry brain can't help but care.

And so it's—Here-I-go-again!
I'll search and search until the end,
to understand things through and through,
'cause learning stuff is what I do.

The solar system is inside a giant bubble that scientists call the *heliosphere*. On the other side of this bubble is the *interstellar medium*, mostly hydrogen and helium gas, that fills up the Milk Way galaxy. *Plasma* (or *solar wind*) is blasted out from the sun in all directions, out past the farthest planets. Eventually the pressure of interstellar medium slows the solar wind to less than the speed of sound. This is the point scientists call *termination shock*, which is the beginning of the end of the heliosphere.

Outer-space distances are gigantic, so astronomers need gigantic units of measure.

For distances within our solar system, they use ***astronomical units*** or **AUs**. One AU is equal to the distance from the Sun to the Earth, or just under 93 million miles. So the distance from the Sun to the beginning of the Oort cloud is 50 thousand AUs.

For the even longer distances between stars and galaxies, astronomers use ***light years***. One light year is equal to the distance a beam of light can travel in one year. Light travels just over 186,000 miles per second, so one light year is 5.8 trillion miles or 63,242 AUs.

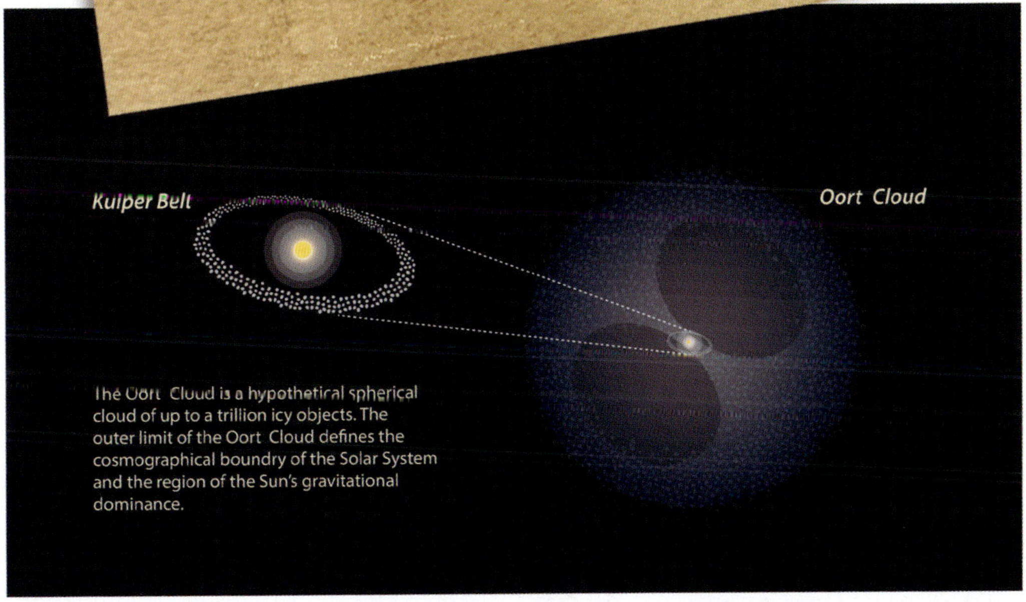

The Oort Cloud is a hypothetical spherical cloud of up to a trillion icy objects. The outer limit of the Oort Cloud defines the cosmographical boundry of the Solar System and the region of the Sun's gravitational dominance.

Inner Heat and Outer Cold

(Chorus)

We like sunshine. We like heat.
Inner planets, small and sweet.
Outer planets? No! No! No!
Big and cold! We will not go!

(Speaker-1)

We start with the Sun, it's one of the stars,
then Mercury, Venus, Earth, and Mars.
All warm and cozy near the Sun.
Inner planets, so much fun.

(Chorus)

We like sunshine. We like heat.
Inner planets, small and sweet.
Outer planets? No! No! No!
Big and cold! We will not go!

(Speaker-2)

Further out, the asteroid belt
is colder than I ever felt.
Oh so cold, it's not so fun
to be so distant from the Sun.

(Chorus)

We like sunshine. We like heat.
Inner planets, small and sweet.
Outer planets? No! No! No!
Big and cold! We will not go!

(Speaker-3)

Jupiter! Saturn! Uranus!
Icy Neptune too!
Outer planets are no place
for folks like me and you.

(All)

Because…
We like sunshine. We like heat.
Inner planets, small and sweet.
Outer planets? No! No! No!
Big and cold! WE…WILL…NOT…GO!

This poem can be a little misleading. It is true that inner planes are generally warmer than the outer planets, but the inner planets can also be very cold Mercury is a good example. It's the closest planet to the Sun so it gets very hot reaching daytime temperatures as high as 900°F. But at night the temperature drops to -300°F, which is much colder than it gets anywhere on Earth. Any idea why this happens? Here are two hints.

Hint 1: Mercury has no atmosphere.

Hint 2: Mercury rotates very slowly. A single Mercury day is equal to 58½ Earth days, which is almost two whole Earth months.

My Lying Eyes

They say it's not true, they say it's a lie,
the Sun doesn't really rise up in the sky.
They say it stands still, and that here on the ground
it's really the Earth that is spinning around.

But I'm standing on Earth, as you all plainly see,
and it sure doesn't feel like it's spinning to me,
because if it were spinning all over the place
then for sure I'd go flying right out into space.

I see what I see, and I know what I know.
The Earth is the center, it's got to be so.
And what everyone says is a bundle of lies,
'cause I know what I see with my very own eyes.

Geocentric Model of our Solar System: the word ***geocentric*** comes from two Greek words: ***gaia*** and ***kentro*** (γαια and κέντρο), which means *Earth-center*.

Heliocentric Model of our Solar System: the word ***heliocentric*** comes from two Greek words: ***helios*** and ***kentro*** (□λιος and κέντρο) which means *Sun-center*.

Sun Center

Geocentric?
Got to go!
Earth at the center?
No! No! No!

Heliocentric?
What a guess!
Sun at the center?
Yes! Yes! Yes!

GEOCENTRIC

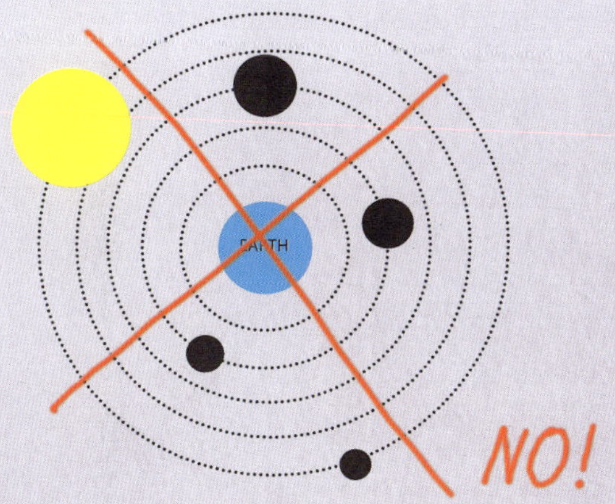

HELIOCENTRIC

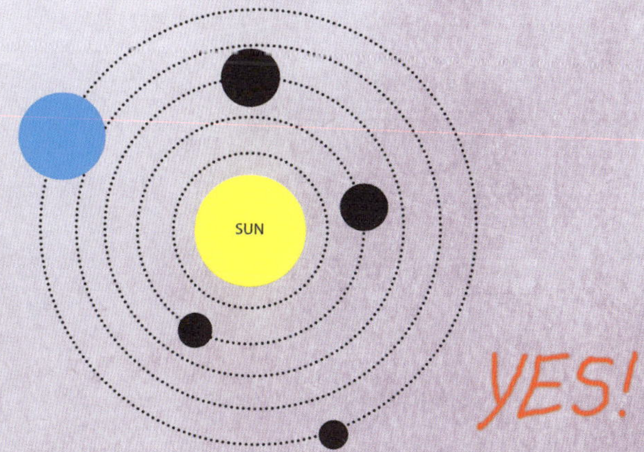

A Superhero Revolution

It's like they were all superheroes.
First comes Aristotle and Ptolemy,
and they were like Batman and Robin.
They came up with this model for the solar system,
and it was very cool
because it could predict the movements of everything in the sky—
the Sun, the Moon, the planets, and all the stars—
but there was just one little problem,
their model had Earth at the center
with everything revolving around it,
which is not the way it is,
but it seemed right
'cause all you had to do was look at the sky,
and sure enough
everything up there seemed to be going in circles around us.

Then a couple thousand years later,
along comes the Fabulous Four—
Copernicus, Kepler, Galileo, and Newton—
and they came up with a model
that was even better at predicting the movements of stuff in the sky,
because their model had the Sun at the center of the solar system,
which was pretty brave because back then
you could get tossed in jail for saying something like that,
but it was true,
and they were superheroes,
and superheroes stand for the truth,
and that's what they did.

So that was it,
a better model for the revolution of heavenly bodies
started a revolution in science,
with more and more people searching for the truth,
and more and more people willing to fight for it.
Superheroes. That's exactly what they were.
Superheroes.

In 1543 Nicolaus Copernicus started what historians call the Scientific Revolution when he published a book that put the Sun at the center of the Solar System. The title of his book is *On the Revolutions of Celestial Spheres*. All scholarly books back then were written in Latin, so here's the Latin title—*De revolutionibus orbium coelestium*.

It's probably one of the most important books ever published. You don't really need to learn the Latin title, but you might have some fun if you do. Imagine you're sitting in class, and your teacher starts talking about Copernicus, and you say, "Yes, Nicolaus Copernicus, *De revolutionibus orbium coelestium*." Either your teacher will be impressed or you'll end up in detention.

Make a Wish

A speck of dust, a tiny stone,
afloat in outer space alone.
Remember this for your exam—
a *meteoroid* is what I am.

Then looking up in dark night skies,
a flash to dazzle human eyes.
Remember this for your exam—
a *meteor* is what I am.

Then every now and then I'm found,
a space rock lying on the ground.
Remember this for your exam—
a *meteorite* is what I am.

But this won't be on your exam,
the magic thing I really am—
a shooting star…one just for you…
a star to make your wish come true.

Most of the meteors we see in the night sky are surprisingly small, between the size of a grain of sand and a piece of gravel. To reach the ground, a meteor has to be about the size of a marble.

Also, you'll probably never get this on a test, but I just learned it myself, and I think it's very cool. Scientists have a name for very large meteors that are extremely bright (as bright as the Moon) and often explode in the atmosphere; they call them ***bolides***. Check it out.

A ***meteoroid*** is a piece of cosmic dust or rock floating around in space. When a meteoroid gets caught in the Earth's gravitational pull it goes faster and faster, eventually entering the atmosphere at speeds somewhere between 25,000 and 160,000 mph. At those speeds there's a lot of friction between the rock and the air molecules, friction makes heat, and the outer layers of the rock begin to burn producing a trail of bright light in the night sky. That's what we call a ***meteor***.

Most meteoroids are too small to gather speed as they enter the atmosphere. Moving at slower speeds, these particles don't get hot enough to burn, which means they go straight from being meteoroids to being ***meteorites*** without ever becoming meteors.

By the way, the word meteor comes from the Greek word ***meteoros*** (μετέωρος) which means "*up in the sky*."

Podkamennaya Tunguska Meteorite

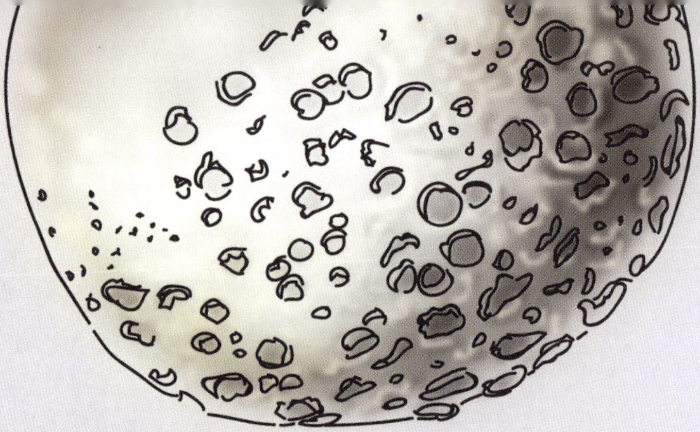

A Nice Place to Visit

I studied the Moon for three whole days,
and now that it's all done,
I'm starting to think that a trip to the Moon
would be only partially fun.

First of all, up there on the Moon,
I could jump like ten feet high.
There's very little gravity
so I'd hardly have to try.

And if I hit a baseball, with
no air to slow it down,
that ball could sail for half a mile
before it hit the ground.

But I'd have to carry a tank full of air,
I'd never be able to run,
and the surface is freezing cold in the dark
and boiling hot in the Sun.

For three whole days I studied the Moon,
and now I have something to say—
I'd really like to visit the Moon,
but I wouldn't like to stay.

How much would a hundred pound person weigh on the Moon and on each of the planets?

(Remember: More mass means stronger gravity and more weight.)

PLANET/MOON	POUNDS
Earth	100
Moon	16.6
Mercury	37.8
Venus	90.7
Mars	37.7
Jupiter	252.8
Saturn	106.4
Uranus	88.9
Neptune	112.5

This poem says that the Moon gets "boiling hot in the Sun," and that is literally true. At sea-level on Earth, water boils at a temperature of 212°F, and the daylight temperature on the Moon gets up to 224°F, so parts of the Moon are "boiling hot in the Sun."

But "boiling hot" is a tricky concept on the Moon. Here's why. On Earth, the pressure of the atmosphere affects the boiling point of water (actually any liquid). The higher your altitude, the lower the atmospheric pressure, and the lower the boiling point of water. So if you went to Alaska and climbed to the top of Mount Denali, you'd be about 20,000 feet above sea-level, and water would start to boil at about 174°F.

But the Moon has no atmosphere and no atmospheric pressure, so water on the Moon would boil at a much lower temperature. As a matter of fact, on the Moon, any water in a liquid state would start to boil immediately. Does that mean that water boiling at say 40°F would make COLD STEAM? I think so. Check out what NASA has learned about steam jets on Enceladus, Saturn'ssixth largest Moon.

When people talk about the phases of the Moon you'll hear them use the words *waxing* and *waning*. Unlike most of the words scientists use, *wax* and *wane* come from neither Latin nor Greek. They are actually two Old English words. ***Wax*** means to *increase or grow*. ***Wane*** means to *decrease or shrink*. So when we say "the Moon is waxing" we mean that each night the lighted surface of the Moon is getting larger. And when we say "the Moon is waning" we mean that each night the lighted surface is getting smaller.

Here are the names for each of the Moon's phase.

- New Moon
- Waxing Crescent Moon
- Waxing Quarter Moon
- Waxing Gibbous Moon
- Full Moon
- Waning Gibbous Moon
- Waning Quarter Moon
- Waning Crescent Moon
- New Moon

Peek-a-boo Moon

Tricky Moon up in the sky,
I look at you and wonder why.
All night long, and daytime too,
I think you're playing peek-a-boo.

Going round and round in space,
showing sections of your face,
sometimes big, sometimes small,
sometimes I can't see you at all.

Tricky Moon up in the sky,
I used to always wonder why,
but now I know, yes now I see,
you're playing peek-a-boo with me.

Turnabout is Fair Play
An Interview with the Moon

A Play In One Act

Cast of Characters:
> ***Talk show host: Isaac Cosmos (Over-the-top enthusiasm)***
> ***Guest: Mariana Moon (Heart-of-Dixie southern accent)***

(Isaac Cosmos, host of the TV show Space—Up Close and Personal, is seated at a desk conducting an interview with Mariana Moon.)

Isaac:

Good evening, ladies and gentlemen,
this is your host Isaac Cosmos,
and welcome to another edition of Space—*Up Close and Personal.*
This evening we have a very special guest.
Love songs by the thousands have been written about her,
she is the force that brings the ocean tides,
once thought to be made of Swiss cheese,
please help me welcome,
the glory of the night sky,
Ms. Mariana Moon.

(Audience applauds, host turns to his guest)

Isaac:

Hello, Ms. Moon,
and welcome to Space—*Up Close and Personal.*

Mariana:

(Smiling mischievously)

Thank you, Isaac.
It's a pleasure to bring a little...glow...to your dreary world.

Isaac:

Already, that famous lunar sense of humor.
So let's get right to it.
What's it like to be Earth's Moon?

Mariana:

Well, Isaac,
you might say that the excitement of being the Moon...wanes and waxes.

(Stopping to see if anyone caught her joke)

You might say that I go through...phases.

(Again stopping)

Isaac:

Phases...? Oh yes, phases!
As in the phases of the Moon.
Aren't you the clever one.
But on a more serious note,
how do you think you compare with other celestial bodies?

Mariana:

There are a lot of celestial bodies out there,
but as far as I'm concerned
there are only three of us that matter—
the Sun, the Earth, and little-ole me.
And yes, I'm the little one in this trio,
so I spend a lot of time in the shadows.

Isaac:

Shadows...
Oh, of course!
You're referring to the phenomenon we Earthlings call a lunar eclipse.
So how does that feel,
to be unceremoniously cast into darkness?

Mariana:

Well, as you know,
most of the time I'm out there doing my thing,
orbiting Earth, minding my own business,
aglow from head to toe in my glorious full moon stage,
when all of a sudden those other two line up just right
so that the Earth comes between me and the Sun,
and there it is, that big shadow right where I'm headed.
I have to admit it makes me a little sad
as mile after mile of my lovely landscape goes dark.
It's not a very nice feeling.

Isaac:

(Deep concern.)

It sounds as though Earth might be engaged
in a bit of astronomical bullying.

Mariana:

No, I wouldn't call her a bully.
Earth and I are in a kind of systematic dance
driven by gravitational forces,
she's bigger than me,
so I end up spending some time in her shadow.
It's just a fact of life.
But it's also a fact that every now and then
I get to return the favor and turn her day into night.

Again, I'm out there orbiting the Earth, minding my own business,
in my new moon stage,
which means you Earthlings can't see me because it's daylight,
when all of a sudden it happens—
my path takes me right between the Sun and the Earth,
and it's HER turn to be in MY shadow.
Now I admit, it's not a very large shadow,
I can't blot her out entirely the way she blots out me,
but as my little shadow moves across her face
it totally blots out the Sun.
The Sun!
Little-ole me blocking out the entire Sun.
Why-if that big ball of blazing hydrogen were hollow
you could fit 64 million of me inside him,
and I block him out.
Please don't think I'm petty,
but I do love it so.

Isaac:

So how do you think Earth feels about that?

Mariana:

Frankly, Isaac, I think she hardly notices.
And besides, I do a lot for her.

Without my gravitational force to slow her down,
one day on Earth would be much shorter,
the tides would be much lower,
and she could forget that nice consistent 23.5 degrees of tilt on her axis.
She'd be spinning every which way,
getting her seasons all mixed up,
and causing all kinds of extreme weather.
So you see, we get along quite well.

Isaac:
Very interesting.
So tell our audience,
what's in the future for Ms. Mariana Moon.

Mariana:
I guess just more of the same.
Being the Moon is not very exciting these days.
There was that time back in the late 1960s and early 1970s
when you Earthlings sent up those Apollo Lunar mission boys to visit me—
kicking up dust with those lunar modules,
planting flags,
traipsing around my surface in lunar rovers,
and leaving all sorts of litter behind,
including a couple of golf balls.

Isaac:
Sounds like a happy time.

Mariana:
Yes it was.
Added a little spice to my world of dusty loneliness.
And who knows, maybe you'll come back someday.
Might even make a little settlement up here.
Then you'd get to see for yourself how it feels
when big-momma Earth blocks out your Sun.

Isaac:
Well, that's it for this week's edition of Space—Up Close and Personal.
Please join us next week
when our guest will be Halley's Comet,
who will explain that famous comet motto—
"Please pay no attention to me, I'm just passing through."

(Audience applauds. The End.)

Solar Eclipse: A solar eclipse happens when the Moon moves directly between Sun and Earth blocking all or part of the Sun from view. A total solar eclipse will happen at some point on Earth on average one and a half times a year, but any particular place on Earth will see a total solar eclipse only every 375 years.

Lunar Eclipse: A Lunar eclipse happens when the Earth moves directly between Sun and Moon casting its shadow over all or part of the Moon. Lunar eclipses happen more often than solar eclipses.

By the way, the word ***lunar*** comes from the Latin word ***luna*** which means *moon*. We find it in the word *lunatic* which is a rude way of referring to a person who is mentally ill. This usage is based on the questionable belief that a full Moon can make people crazy.

Ring Around the Sun

When morning sun brings break of day,
and stars begin to fade away,
amid dawn's rosy reverie
the Morning Star's the last I see.
    Ring around the Sun-Sun,
    having lots of fun-fun.
    Morning Star! Morning Star!
    Who am I?

When evening sun is in decline,
and stars of night prepare to shine,
then from the west in bright delight
the Evening Star's the first in sight.
    Ring around the Sun-Sun,
    having lots of fun-fun.
    Evening Star! Evening Star!
    Who am I?

So here's a secret you should know
about these two that come and go.
They're both one planet having fun
while making rings around the Sun.
    Ring around the Sun-Sun,
    having lots of fun-fun.
    I'm a planet, not a star.
    Who am I?

Planet Venus. Yes, it's true.
Morning Star and Evening, too.
Lovely Venus having fun,
playing Ring-Around the Sun.
    Ring around the Sun-Sun,
    having lots of fun-fun.
    Morning Star! Evening Star!
    Both are me!

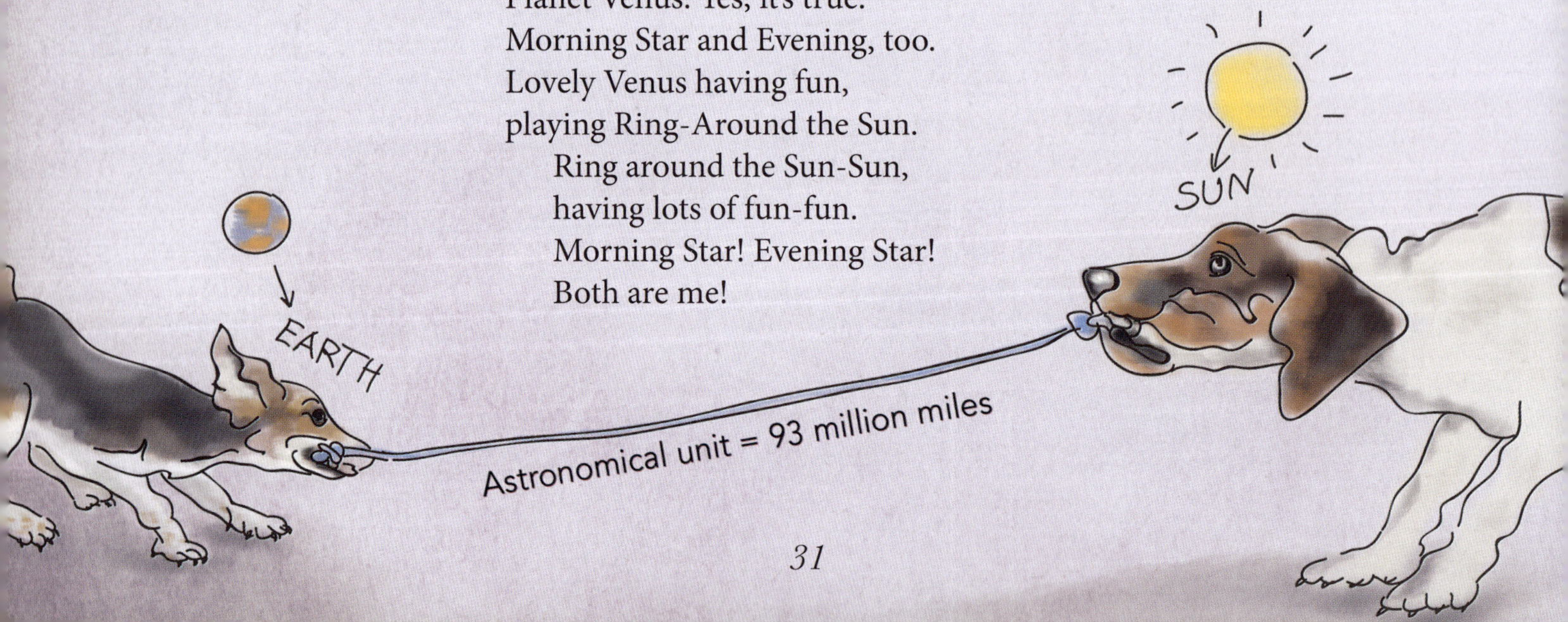

Except for the Sun and the Moon, Venus is the brightest object in the sky. The orbit of Venus is closer to the Sun than the orbit of Earth, which is why, when viewed from Earth, Venus is always seen close to the sun—in the morning when the Sun rises in the east, and in the evening when the Sun sets in the west.

Transit of Venus

That little shadow is Venus moving across the face of the sun. It's called a *transit of Venus*, and they come in pairs, each eight years apart. The last transit of Venus occurred in 2004 and 2012. The next pair will be in 2117 and 2125.

In advance of the transits of 1761-1769, scientists traveled all over the world to measure the movement of Venus across the sun's face. Then in 1771, using those measurements, astronomer Jérôme Lalande was able to make a pretty accurate calculation of the distance from the Earth to the Sun.

Do you remember what that distance is and how we use it? Yes! The distance is 93 million miles, and we call it an astronomical unit, which is the unit of measure we use to express distances within our own solar system.

Sunshine Science

From 93 million miles away
you turn my nighttime into day.

Hydrogen fusion! That's your power!
Burning hour after hour!

Sunspot circles on your face!
Solar flares shoot out in space!

Solar wind! Magnetic swarm!
A giant electronic storm!

Solar science! So much fun!
Learning all about the Sun!

From 93 million miles away,
you turn my nighttime into day.

The sun is a ball of gas and plasma. (We usually talk about the three phases of matter, but there are actually five—solid, liquid, gas, plasma, and Bose-Einstein condensates.)

The power of the sun (and all stars) is generated by what scientists call the **Hydrogen Fusion Process** in which hydrogen atoms turn into helium atoms. Here's how it works. The nucleus of a hydrogen atom has one proton. The nucleus of a helium atom has two protons and two neutrons. Gravitational forces in the sun's core create humongous pressure and temperature which causes four hydrogen atoms to breakup and become one helium atom.

During this conversion, the hydrogen atoms lose a little of their mass which is converted into energy (MORE ENERGY THAN IT TOOK TO CAUSE THE FUSION). That energy travels through a layer of the sun called the radiative zone (which takes millions of years) and is eventually released as what scientists call electro magnetic radiation, or what you and I call light.

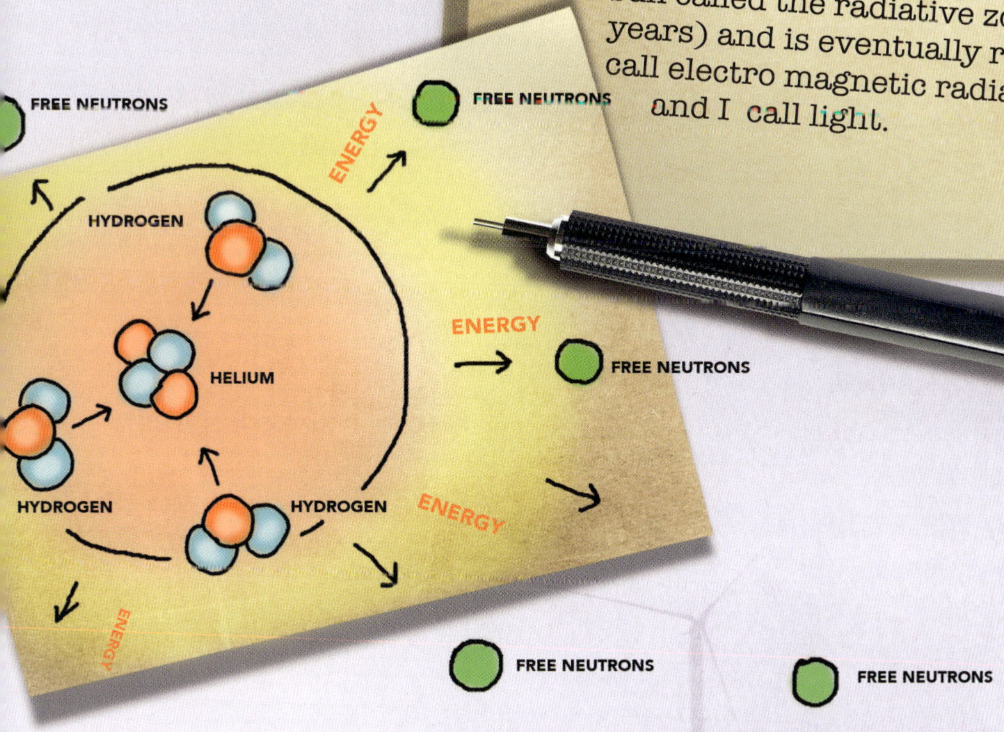

A Nerd Love Poem

My name is Roshanda Hale,
and for the first time in my life
a boy has written me a love poem,
and that boy is none other than Nerd-One
himself, Alvin Lofton.
Just listen to what that fool wrote.

STAR FIRE

TO ROSHANDA HALE

I am intrigued by the life of stars:
How an aging red giant looses its outer layers
to become a denser, cooler white dwarf.

Or how supergiants go nova
in an explosion as bright as a whole galaxy,
then just as suddenly collapse to form black holes.

And the celestial dance of binary stars,
bound to each other in balanced rotation
round a common center of mass.

And I have found it difficult to comprehend
the reality of stars—
these enigmatic manipulators of time,
space benders,
cosmic furnaces ablaze in the boundless black of space.

Yet last night,
when in the constellation Orion
I beheld the blue brightness of Rigel,
my thoughts were consumed,
not by the mystery of stars,
but by the fire that burns in the crucible of my heart
when you look at me and smile.

Star light, star bright,
brightest star I see tonight,
I wish I may, I wish I might,
dream a dream of you tonight.

by Alvin Lofton

So there it is,
my first love poem,
it's from Alvin Lofton,
and have only one thing to say—
THAT NERD HAS GOT IT GOING-ON!

Why do you think this poem refers to stars as "these enigmatic manipulators of space time?" You may know this already, but here are a some cool facts about how long it takes light from stars to reach us on Earth.

Close: It takes about eight minutes for light to travel the 93 million miles from the Sun to us on Earth. (As you know, the Sun is actually a star.)

Far: Other than the Sun, the nearest star to Earth is Alpha Centauri (a triple-star system) which is 4.37 light-years away. That means that when you look at Alpha Centauri the light you see started its journey toward Earth over four years ago, so you are looking at the Alpha Centauri as it was four years in the past, not as it is now.

Farther: There's a star in the constellation Cassiopeia that is the farthest star we can see with the naked eye (without a telescope). It's Over 16,000 light years away, which means that the light we see today left that star 12,000 years before the Egyptians started building the Great Pyramid of Giza.

Farthest: The farthest individual star we have seen so far is over 9 billion light years away. Our solar system is only 4.5 billion years old. That means that when the light we see today left that star 9 billion years ago, our solar system (Earth and all) did not even exist.

By the way, our ability to see an individual star that far away was the result of a phenomena called ***gravitational lensing***. This is what happens when light headed our way is bent as it passes through the *gravitational field* of a gigantic mass, in this case a galaxy cluster. Without gravitational lensing, we can only see individual stars that are within a distance of 100,000 light years.

Black Hole Brain

(In the voice of Shayna Potts—Goth 7th Grader)

It happened again.
In the middle of science class
Mr. Grindle was talking about black holes,
how they're all over the Universe,
but you can't see them,
because they are so massive with so much gravity,
they suck up everything around them,
they even suck up light,
and I started thinking:
If only I had my very own black hole
I'd never have to rake the lawn again.
I'd just toss that hole out on the grass
and the leaves would all be gone.
And the fun I'd have at school!
I'd roll a tiny black hole to the front of class,
and watch it suck up Mr. Grindle
and smush him into a singularity.

Which is when I must have laughed out loud,
cause the next thing I heard was Mr. Grindle:
"Well now, Miss Potts,
if it's that funny, please,
don't be selfish,
share it with the rest of us."

It's happening more and more,
almost every day,
My wild imagination simply
steals my brain away.
"Black holes are mostly far away,
but some are close," I said.
"I know because I think I have one
stuck inside my head."

This whole star life-cycle thing is very cool. For example, when a star 4 to 8 times the size of the Sun collapses it becomes a neutron star. Neutron stars have so much gravity that they can suck up nearby stars, and as they gain more and more mass they get smaller and smaller. HA!

A black hole forms when a bigger star, at least 8 times the size of our Sun, reaches the end of its life cycle. There's a big flash as the atoms collapse and form a mass so dense and with so much gravity that nothing can escape it, not even light, which is why we call it a black hole.

Here's my favorite thing about black holes: If planet Earth were compressed to the density of a black hole, how big would it be? (Are you ready?) It would be the size of a peanut. A PEANUT! The whole Earth—mountains, continents, oceans, your school—all of it would fit inside a peanut shell. Scientists are learning lots of new stuff about black holes. We now know, for example, that there's a black hole at the center of most galaxies, it's the mass around which galaxies rotate.

This might be an area you'll want to learn more about. You might even decide to become an expert. I think I'd like to see a movie about a seven-year-old girl who becomes a black hole expert and saves the world.

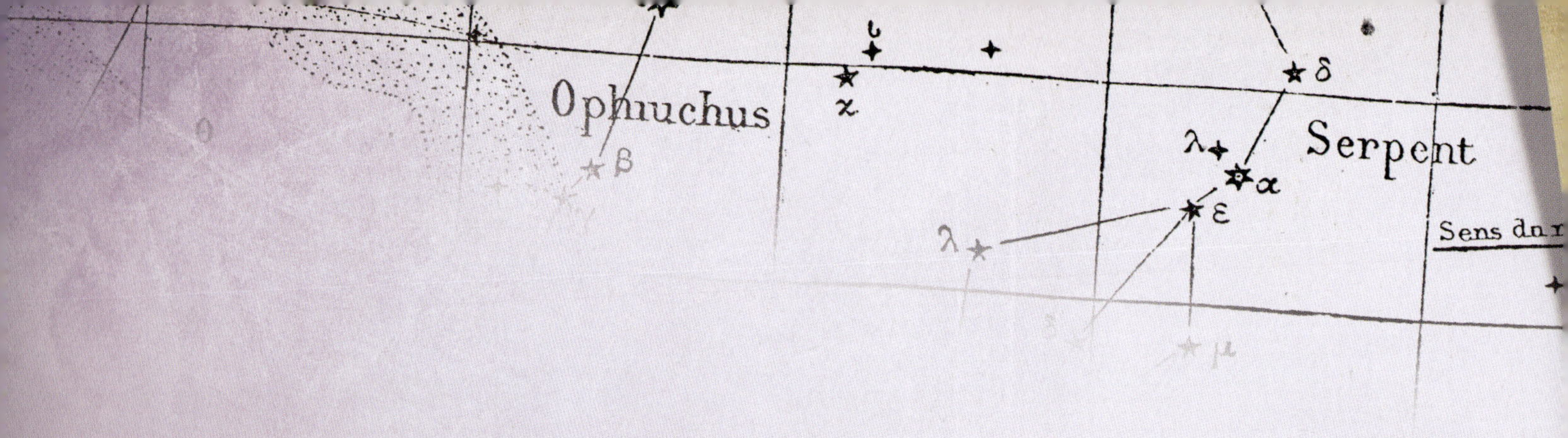

Biggest?

The biggest thing you'll ever see?
It's got to be a galaxy.
These giant cosmic reservoirs,
each with a hundred billion stars.

All the stars now known to science —
neutron stars, white dwarfs, red giants,
pulsars, yes, and black holes too
so dense not even light gets through.

Cosmic clouds and asteroids
floating in the cosmic voids.
Billions of planets, kazillions of moons,
floating in space like enchanted balloons.

Humongous cosmic reservoirs
each with a hundred billion stars.
The biggest thing you'll ever see?
It's GOT to be a galaxy.

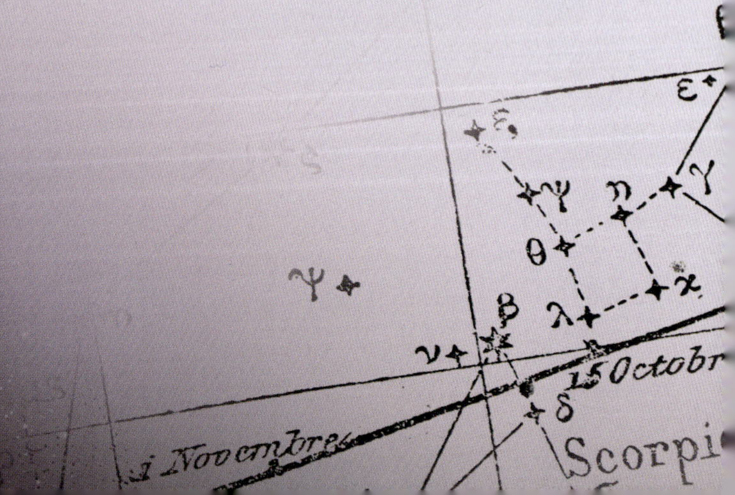

Is a galaxy really the biggest structure in the Universe? Kind of. It's the biggest individual structure, but galaxies cluster together to form even larger identifiable groups. The Milky Way, which is our galaxy, consists of about 200 billion stars, which is big, but the Milky Way itself is part of an even bigger structure known as the Local Group, which is composed of 54 galaxies. That's even bigger, but once again the Local Group is part of an even bigger structure called the Virgo Supercluster, which contains over 100,000 galaxies.

As big as the cosmos is, that's how much stuff there is to learn. Big stuff and little stuff—we never run out of cool things to learn. And let me say it again, learning is not just something you do to get good grades and get a job. Learning can be something you do because it's fun. I had a friend in fourth grade who could rattle off the batting averages of hundreds of Major League baseball players. Nobody made him learn that, for him it was fun.

The Wall

I look at the stars on a winter night,
each one in its proper place,
and wonder—*how can it possibly be*
that there is no end to space?

There must be an end, a giant wall—
I finally decide.
Then I wonder again—*if there is such a wall,*
then what's on the other side?

Hubble Ultra Deep Field, NASA

In this little poem I'm having fun with the logical paradox inherent in the concept of endlessness. I'm going to use it as an opportunity to talk to you about one of my favorite things in recent astronomy—a photograph recorded in December of 1995, known as **Hubble Deep Field**.

Here's what happened. Scientists selected a tiny area of the sky near the Big Dipper. If you look there with your naked eye, you see nothing, total blackness. Scientists then pointed the Hubble Space Telescope at this little spot of nothing and what they got was an image containing almost 3,000 objects. But here's my favorite part—most of those objects, those bright spots you see in this photograph, are not stars, they're galaxies, which means each little smudge of light in this picture contains hundreds of billions of stars.

In 2004 with *Hubble Ultra-Deep Field*, and again in 2012 with Hubble *Extreme Deep Field*, using similar techniques, scientists recorded images even deeper into space, and the results were always the same. In whatever direction we look, we find galaxies, lots and lots and lots of galaxies.

Alien Eyes?

Our Universe is a zillions stars
afloat in an endless sea,
expanding like a drop of cream
in a giant cup of tea.

So many stars in that cosmic cup
there almost has to be
some other eyes that search their skies
and wonder about me.

Exoplanets are planets that orbit stars outside our solar system. In the ongoing search for exoplanets, scientists are particularly interested in the ones that might be able to support life, which are those that lie within the *circumstellar habitable zone*.

That's the term scientists began to use in 1993 to describe exoplanets that are about the same size as Earth, orbiting a star about the same size as our Sun at about the same orbital distance. (Scientists used to use the informal term *Goldilocks zone* to refer to an area around a star where the temperature is "just right" for water to stay in liquid form.)

Have you ever wondered if there might be extraterrestrial life out there in the Universe? Here's something to think about.

1: In 2013, NASA astronomers estimated that in the Milky Way alone there could be as many as 11 billion Earth-sized planets orbiting Sun-like stars. That's 11 BILLION in the Milky Way Galaxy alone!

2: NASA astronomers also estimate that there are 100-200 billion galaxies, and expect that estimate to increase dramatically as we get even more powerful telescopes.

Put these together and what do you get? That in each of the hundreds of billions of galaxies there are billions of potentially life-supporting exoplanets. So, extraterrestrial life? What do you think?

For most of our existence on Earth, we humans lived with the idea that we were at the center of it all, that the sun and the Moon and the planets and all the stars revolved around us. Then, starting about 500 years ago, all that began to change. The collective work of Copernicus, than Kepler and Galileo, and finally Newton proved that we humans were not at the center, but that our Earth was just one of several planets orbiting the sun. Since then, as we have become more and more aware of the immensity of the universe, the physical significance of our small planet has continued to diminish. During the same time, human population increased from 500,000 in the year 1500 to an estimated eight billion by the year 2024.

In the face of all that change, some people feel that their lives are unimportant. Well, that is no longer a problem for the voice speaking in this poem. She has decided that in this big-ole universe she as an individual is important—"what I think matters, and what I do matters, and what I stand for matters, and that I matter." So I wrote this poem because you matter too, and I want you to know it.

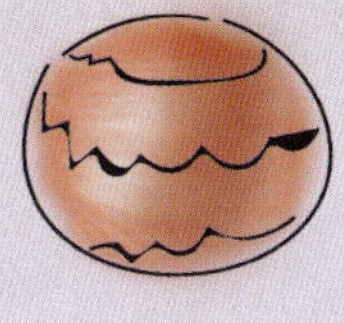

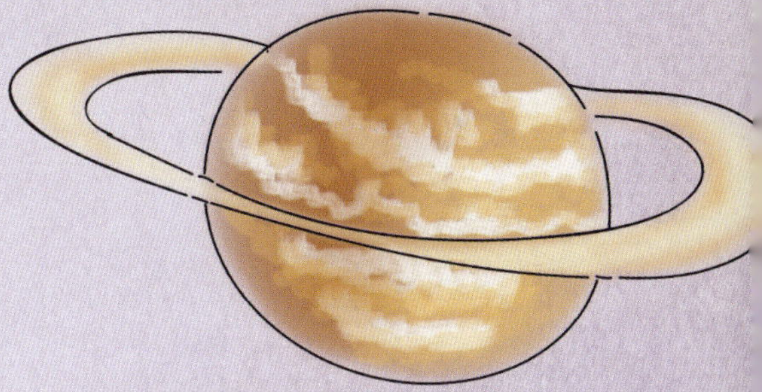

My Home in the Universe

I am me.
I live in Apartment 2B,
at 2829 Paris Ave,
in the City of New Orleans,
State of Louisiana,
in the United States of America,
on the continent of North America,
on the surface of the Earth
a small watery planet in the third orbit around the Sun,
which is a medium size star on the Orion Spur of the Milky Way,
our very own spiral galaxy of over 200 billion stars,
which together with three other large galaxies and dozens of small ones
form The Local Group,
one of a hundred other local groups,
each consisting of 10 to 10,000 galaxies,
which together form the Virgo Supercluster,
itself located in one lobe of the even larger Laniakea Supercluster,
one of a vast web of superclusters that comprise what we call
the Known Universe,
and, having learned all this,
I am perfectly clear
that I am not the center of the universe,
yet I have arrived at the conclusion that what I think matters,
and what I do matters,
and what I stand for matters,
and that I matter,
because I…am…ME,
and this…is where I live,
my home in the universe.

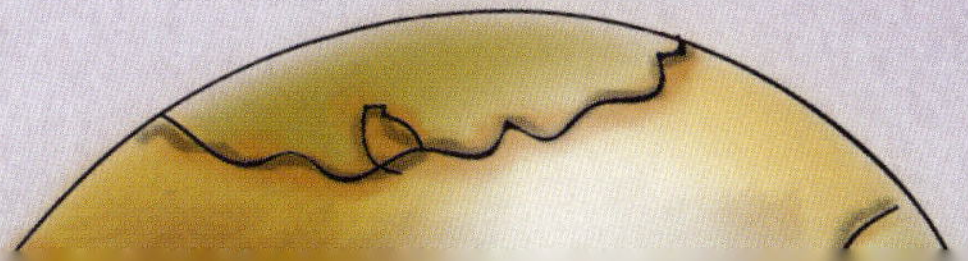

You Will Find Me
The Voice of an Unnamed Star

You will find me
past the prime meridian
at eighteen hours seven minutes four.

You will find me
beneath the celestial equator
by twenty-three degrees, forty-five minutes,
and not one second more.

Ruled by Jupiter,
the warm and wise,
I am but a sparkle in the eye of the Archer,
a speck in the cloud of the Milky Way,
so small beside these giants of the Universe,
and yet...
a line drawn through me from the center of the Sun
will pierce the heart of a galaxy.
And so,
in this ocean of eternal night,
I make day.
I burn to give this light away
and in burning earn the right to say—
I am a star.

So, in the twilight of some cool September evening,
remember me,
and perceiving how I loved you,
look for me again,
just above the southern horizon,
for I will be there,
and you will find me.

Here are four things that will help you understand this poem.
1. Only a few of the really big stars have names. Like the little star in this poem, all the rest are designated only by a number.
2. We use longitude and latitude to locate the position of objects in the sky, just as we do on Earth. The coordinates of this unnamed star indicate that it is located in the constellation Sagittarius, "the archer." And although it has no scientific significance, in astrology, Sagittarius is said to be "ruled by Jupiter."

3. On a clear night in a southern autumn sky, if you look beyond the constellation Sagittarius you will see a cloud of stars called the Milky Way, which is actually our own galaxy.
4. And most important, there are two kinds of things in the universe—those that burn to make light, and those that do not.

An Invitation

Some years ago it became my life's dream to write a bunch of poems about science, poems that would be both fun and instructional, poems to simultaneously entertain you and help you become life-long learners.

I've been at it now for over twenty years, and I have begun to understand that this little book and the ones to follow, are just a first step, and that the next step depends on you.

That's why I am inviting you to join me in the wonderful work of awakening the joy of learning in the hearts of your generation. We humans are proud of our tall buildings and our two-thousand-year-old aqueducts and our four-thousand-year-old pyramids, and we should be. But what we can be most proud of is the body of knowledge we have accumulated over the millennia, each generation adding its part. I invite you to help keep that process healthy and strong in your generation.

Look back through the pages of this book, pick-out a piece that you really like, then do something with it and send it to me. You might make a video of you and your friends performing it; or create your own original illustration or animation; or write your own original poem or play; or simply write your thoughts after reading it; or maybe something else I can't even imagine. As time goes by we at *Brod Bagert's HeART of Science* will be selecting some of your material and (with your permission) sharing it with others just like you.

I hope that many of you will accept this invitation and have fun doing it. There may even be a few of you who will like it a lot, a few of you for whom it will become the joy of your lives, as it has been in mine. Maybe you'll be the ones to help take this HeART of Science concept to a whole new level. Maybe, eventually, you'll be the stars that bring light to the night skies of your generation. Please accept this as your invitation.

Dedication

To my wonderful grandson, Julian.
B.B.

To my Dad, Anatoliy.
N.K.

Publishing Information

Published by Living Road Press
San Antonio, TX

ISBN 978-1732151536

Acknowledgement

A big thanks to Eric M. Schlegel, Ph.D. Vaughan Family Professor, Physic & Astronomy, Astrophysicist at the University of Texas at San Antonio for his explanation of the often misunderstood concept of "Zero G".

Thank you to Ellen White, Micronaut Specialist at the Scobee Education Center, San Antonio College, for reading this manuscript and for sharing her observations about the importance of instilling curiosity in children.